1

ONE looks like my one finger.

KOW

Waxay u eegtahay fartayda kowaad.

starting from Number One!

Waxaan ka bilaabi doonaa tirade kowaad!

Say-along our little jingle

Nagula hees, sheekadayada yar!

Want to learn our number names?

Ma jeclaan lahayd inaad barato magacyada lambarrada?

It is very easy and a lot of fun!

Way fududahay xiiso bandanna way leedahay!

Copyright © 2018 by Jieeun Woo
Illustrations © Jieeun Woo

Cover by | Lumpy Publishing
Layout by | Lumpy Publishing
Translated by Muhyadin Dayib
Coloring by Jieeun Woo and Maria Mirabella

All rights reserved. No part of this book may be reproduced or transmitted in any form or by any means whatsoever, including photocopying, recording or by any information storage and retrieval system, without written permission from the publisher and/or author: missanna@missannabooks.com.

Library of Congress Control Number: 2018902040

Names: Miss Anna, author.
Title: Number story : numbers teach children their number names / Miss Anna.
Description: Portland, OR: Lumpy Publishing, 2018.
Identifiers: ISBN 978-1-949320-12-1| LCCN 2018902040
Summary: The pictures and rhymes present stories which introduce numbers 0-10.
Subjects: LCSH Numeration—English—Somali--Pictorial works--Juvenile literature. | BISAC JUVENILE NONFICTION /
Languages: English—Somali
Classification: LCC QA141.3 .M57 2018 | DDC 513—dc23

Publisher: Lumpy Publishing
Website: www.missannabooks.com
Email: missanna@missannabooks.com

Paperback: ISBN 978-1-949320-12-1
Printed in the U.S.A.   1 3 5 7 9 10 8 6 4 2

SHEEKADA TIRADA

THE NUMBER STORY

SMALL BOOK ONE

ENGLISH – SOMALI

Numbers Teach Children
Their Number Names

written and illustrated by

MISS ANNA

Early Reader Edition of *The Number Story 1*
Bronze Medal Winner, 2016 Wishing Shelf Book Award

LUMPY PUBLISHING
OREGON

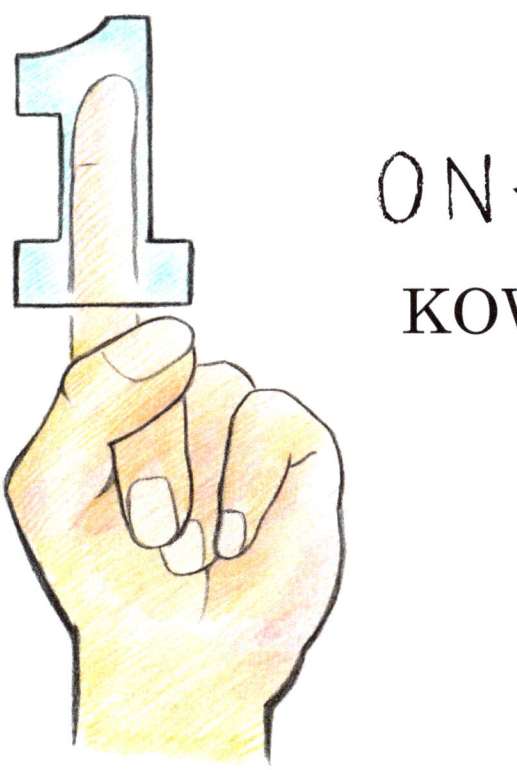

2

TWO trails a tail.

LABA

Waxay calaamadisaa dubka.

A TAIL! DUB!

3

THREE has bumps.

SADDEX

Waxay leedahay qaloocyo.

BUMPY! QALOOC!

4

FOUR carries a sail.

AFAR

Wuxuu qaadaa badmaax.

5

FIVE is a racing track.

SHAN

Waa baabuur tartamaya.

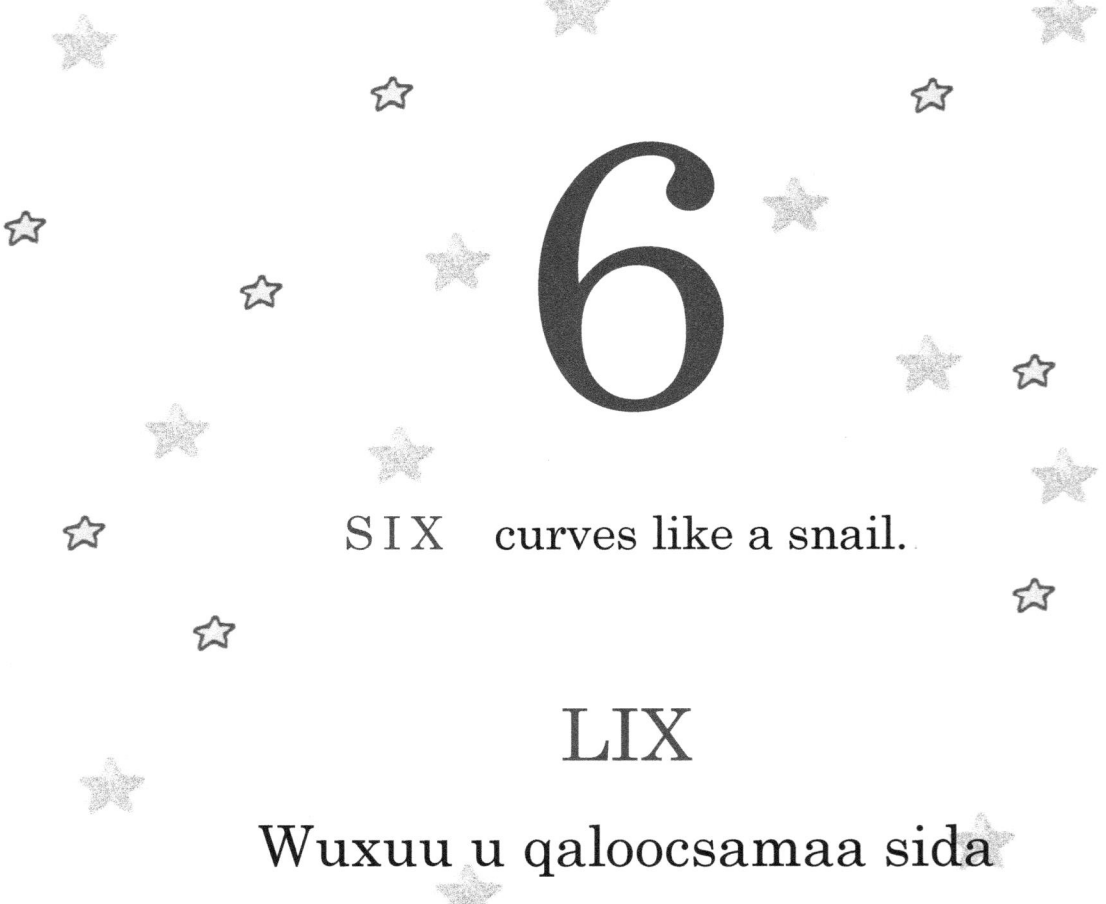

SIX curves like a snail.

LIX

Wuxuu u qaloocsamaa sida masaabiirta.

A SNAIL! MASAANIIR!

7

SEVEN has a sharp angle.

TODDOBA

Wuxuu leeyahay xagal fiiqan.

BE CAREFUL! IT'S SHARP!

ISKA JIR! WAA SEEF!

8

EIGHT is a rollercoaster rails.

SIDDEED

Waa khaanado isku wareegsan.

NINE is a bubble on a stick.

SAGAAL

Waa xumbbo la isku dhejiyay.

A BUBBLE!

XUMMBO!

10

T E N is an eye of a whale.

TOBAN

Waa il kaliya oo bad weyni leedahay.

WINK!

CAWAR!

HELLO! NABADEEY!

And
Iyo

0

ZERO is an empty pail.

EBER

Waa baaldi madhan.

Thank you for playing with us today.

We had a lot of fun too!

Waad ku mahadsantahay inaad

nala ciyaarto maanta.

Xaraabaad badan ayaan yeelanay sidoo kale!

We are your Number friends,
Zero to Ten,
Who will be here for you~
Waxaanahay tirada saaxiibadaa
Eber ilaa Toban.
Had iyo jeer waan kuu joogi doonnaa.

Bye-bye now!
See you again soon!
Babadeey-nabayeey hadda!
Is arag danbe!

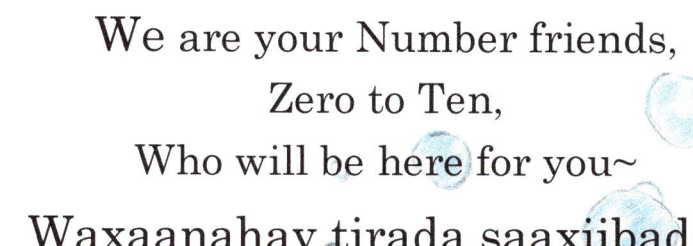

The Numbers are *SINGING* too!

To *sing-a-long*, look for Miss Anna Number Story at your favorite music store like iTUNES.

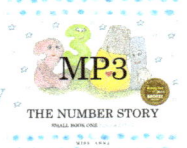

The *Read-to-Me* Editions

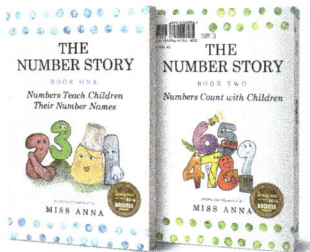

Number Story 1 & 2
ISBN: 978-0-996216-48-7

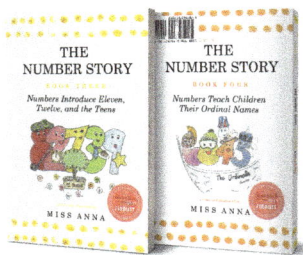

Number Story 3 & 4
ISBN: 978-1-945977-01-5

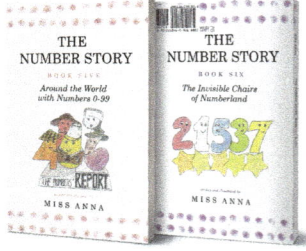

Number Story 5 & 6
ISBN: 978-1-945977-06-0

For more Miss Anna books to love,
visit us at

www.missannabooks.com

Numbers are working hard all over the *world*!

Come Travel the World with Us!

CPSIA information can be obtained
at www.ICGtesting.com
Printed in the USA
LVHW070006091019
633643LV00012B/52/P